BASIC ICT SKILLS & SHORTCUT KEYS

Er.Rohit Kataria

First Published in November 2021

ISBN: 978-93-5472-857-0

BLUEROSE PUBLISHERS

www.bluerosepublishers.com

info@bluerosepublishers.com

+91 8882 898 898

Cover Design:

Aveek

Typographic Design:

Namrata Saini

Distributed by: BlueRose, Amazon, Flipkart, Shopclues

Preface

Computers play vital role in today's world. World looks incomplete without computers.

The Aim of writing this book is to help those who wish to learn Basic ICT skills and keyboard shortcuts for various applications, which makes working on computers as cakewalk. This book contains Basic ICT Skills & lot of Shortcut Keys based on MS-Office (Word, Excel, and PowerPoint), Adobe Photoshop, Adobe PageMaker, Corel PaintShop Pro, Corel Draw, Chrome Browser and Typing Skills with which one can use such software's easy & quickly and can have quick results in one's work.

As a IT Faculty, I observed in these days of technology that every person is using the computer who is part of any school/institution/office, but major concern is that they are not taking advantage of various tact's to use it, which can save their valuable time and they can do work very efficiently as professional. This thought took place in my mind when one day a student in IT lab could not complete the given task in stipulated time and failed to submit the task, I observed that a student should use some keyboard shortcuts to complete such task on time. Then I decided to write it, which could solve such problem in present & future. I really thankful to that student due to whom I got the idea of writing this book.

I am sure that this book will save your valuable time to use computer software's with keyboard shortcuts and make you expert in the field of computer/IT.

Contents

Introduction to Computer

Computer is part of our life; as we know very well every person who belongs to any work or position in society is depend on it. Now days it is very difficult to complete your daily work without computer. So there are so many numbers of tools and shortcuts to work on it, which will make your job (work) much easier and faster.

As we all know keyboard plays an important role in computer/laptop. But 70% of the people are using it only for input data, not for quick use of applications. So first of all you should learn the quick way of use applications. Many shortcut keys & tips are introduced in this book for it.

A computer is an electronic device that manipulates data according to a set of instructions. A computer can use many programs to carry out a wide range of useful tasks. It is programmed to carry out sequence of arithmetic and logical operations automatically.

A broad range of industrial and consumer products use computers as control systems.

Early computers were meant to be used only for calculations. Simple manual instruments like the abacus have aided people in doing calculations since ancient times.

Operating System

An Operating System **(OS)** is software program that manages the hardware and software resources of a computer. A key component of system software, The OS performs basic tasks, such as controlling and allocating memory, organizing the processing of instructions, controlling input and output devices make easy networking and managing files.

In short words Operating System is like a soul of our computer system, as soul makes the physical body alive. In the same way Operating System is so important for computer.

Some of the major facilities provided by a modern operating system are:

- Easy interaction b/w humans and computers.
- Starting computer automatically when power is turned on.
- Loading and scheduling users programs along with necessary compilers.
- Controlling input and output.
- Controlling program execution.
- Scheduling processes.
- Managing use of main memory.
- Managing and manipulating files.
- Providing security to users jobs and files.

- Accounting resource usage.

Common operating system tasks:-

- Monitoring performance
- Correcting errors
- Providing and maintaining the user interface
- Starting the computer
- Reading programs into memory
- Managing memory allocation to those programs
- Placing files and programs in secondary storage
- Creating and maintaining directories or folders
- Formatting diskettes(disks)
- Controlling the computer monitor
- Sending jobs to the printer
- Maintaining security and limiting access
- Locating files
- Detecting viruses
- Compressing data

Files and Folders in Windows

File

A file contains a part of information stored in the computer. There are many types of files, like text files, audio files, image files etc. The file has a filename and an extension (called format of file).

Folder

A folder can hold a collection of different types of files as different type of clothes in suitcase as well as it can contain other folders in it.

Measures to protect computer

Setting up users and passwords: By setting up a list of authorized users on computer, we can limit how users can enter the computer. To access data on the computer, the user must log in with his/her username and the corresponding passwords. Anybody without the knowledge of the password is not allowed into the system.

Use of protective software: It is advisable to use software that will scan the drives or downloads for presence of suspicious software in them. These software scan files, whether there could be disbelieving material part of any file. Antivirus software is one such software for Example: -McAfee, Symantec, Kaspersky, Norton, Quick Heal etc.

Using secure passwords: As passwords are very important to a computer user. Password is the only knowledge based on which one can enter either the computer or one's inbox (or mailbox). You should ensure that your password is not easily guessable. It should not be a dictionary word and also must contain a mix of different types of characters from keyboard (like numbers, special characters and alphabet).

Lock your computer: It is very necessary to lock your account, when the computer is left unwatched. This way no one can enter into your account, if you are away from the computer for a short while. The key combination that will allow locking the computer depends on the particular operating system.

Hardware safety and security

The physical components of a computer system are called hardware. Hardware devices can be seen and touched.CPU, Monitor, Printer, Scanner, Hard Drive, CD's ROM, Modem, Port and Cables are all hardware devices of a computer system.

To maintain the hardware security and safety, you must follow some guidelines as given below:

1. **Computers Must be Kept Clean:** There must be regular cleaning of dust and debris, and also we should avoid eating and drinking near the computer.
2. **Computers should be Kept at Moderate Temperature:** High or low temperature may damage motherboard's memory and disk drives memory.
3. **All Computers Should be Connected to Surge Protectors:** The surge protectors will prevent voltage surges from damaging the system.
4. **Computers and Disks Must be Kept Clear of Static Electricity:** To avoid zapping your computer

components (like computer memory and hardware) with static electricity other precaution is to ground the static electricity before touching any of the internal components.

5. **Aware of Keyloggers:** A hardware keylogger is a small piece of hardware that is manually attached to the keyboard wiring plugged into the PC's CPU. It is placed at an intermediate position between the CPU plug and keyboard's wire. It captures all your keyboard stokes while typing. You need not worry about keylogger for your personal computer until a close one uses this.

Utility tools of Windows

➢ **Disk Cleanup:**

It is part of system tools in windows which helps you to remove the unnecessary files from your computer hard disk to free up hard disk space. It removes all the downloaded programs and other files which are not being used for a long time. It also empties the Recycle bin.

To use Disk Cleanup follow the given steps:

1. Click the **Start** ⟶ **All programs** ⟶ **Accessories** ⟶ **System Tools** ⟶ **Disk Cleanup.**
2. **Disk Cleanup: Drive Selection** then click on drop down arrow and click **OK.**

Otherwise directly search disk cleanup option in search box.

- **Disk Defragmenter**

It helps you to rearrange the fragmented data so that your hard disk drives can work more efficiently. Disk defragmenter runs on a schedule, but you can also defragment your disks manually.

To use Disk Defragmenter follow the given steps:

1. Click the **Start** ⟶ **All programs** ⟶ **Accessories** ⟶ **System Tools** ⟶ **Disk Defragmenter.**
2. The Disk Defragmenter window appears on the screen.
3. Select the disk you want to defragment Under Current Status.
4. You may check the percentage of fragmentation on disk in the **Last Run.**
5. Click the **Defragment disk.**

- **Antivirus Software**

It is program that detects, prevents from malicious programs such as viruses. You can protect your system against viruses by using Antivirus software. There are many antivirus programs available in market or online web store.

Windows 10 Shortcut Keys

Shortcut		Action
Windows key	→	Open or close Start Menu.
Windows key + A	→	Open Action center.
Windows key + C	→	Open Cortana in listening mode.
Windows key + D	→	Display or hide the desktop.
Windows key + E	→	Open File Explorer.
Windows key + G	→	Open Game bar when a game is open.
Windows key + H	→	Open the Share charm.
Windows key + I	→	Open Settings.
Windows key + K	→	Open the Connect quick action.
Windows key + L	→	Lock your PC or switch accounts.
Windows key + M	→	Minimize all windows.
Windows key + R	→	Open Run dialog box.
Windows key + S	→	Open Search.
Windows key + U	→	Open Ease of Access Center.

Windows key + X	→	Open Quick Link menu.
Windows key + Number	→	Open the app pinned to the taskbar in the position indicated by the number.
Windows key + Left arrow key	→	Snap app windows left.
Windows key + Right arrow key	→	Snap app windows right.
Windows key + Up arrow key	→	Maximize app windows.
Windows key + Down arrow key	→	Minimize app windows.
Windows key + Comma	→	Temporarily peek at the desktop.
Windows key + Ctrl + D	→	Add a virtual desktop.
Windows key + Ctrl + Left or Right arrow	→	Switch between virtual desktops.
Windows key + Ctrl + F4	→	Close current virtual desktop.
Windows key + Ctrl+Enter	→	Open Narrator.
Windows key + Home	→	Minimize all but the active desktop window (restores all windows on second stroke).
Windows key + PrtScn	→	Capture a screenshot and save in Screenshots folder.

Windows key + Shift + Up arrow	→	Stretch the desktop window to the top and bottom of the screen.
Windows key + Tab	→	Open Task view.
Windows key + "+" key	→	Zoom in using the magnifier.
Windows key + "-" key	→	Zoom out using the magnifier.
Ctrl + Shift + Esc	→	Open Task Manager.
Alt + Tab	→	Switch between open apps.
Alt + Page Up	→	Move up one screen.
Alt + Page down	→	Move down one screen.
Ctrl + Alt +Tab	→	View open apps
Ctrl + C	→	Copy selected items to clipboard.
Ctrl + X	→	Cut selected items.
Ctrl + V	→	Paste content from clipboard.
Ctrl + A	→	Select all content.
Ctrl + Z	→	Undo an action.
Ctrl + Y	→	Redo an action.
Ctrl + D	→	Delete the selected item and move it to the Recycle Bin.
Ctrl + Esc	→	Open the Start Menu.

Ctrl + Shift + click a taskbar button	→	Open an app as an administrator.
Shift + right-click a taskbar button	→	Show the window menu for app.
Ctrl + F4	→	Close the active window.
Windows key + I	→	Open Settings.

MS-Office

Microsoft Office, or simply Office, is a family of client software, server software, and services developed by Microsoft. It was first announced by Bill Gates on August 1, 1988, at COMDEX in Las Vegas.

Office is produced in several versions targeted towards different end-users and computing environments.

Some Core apps and services

- Microsoft Word
- Microsoft Excel
- Microsoft PowerPoint
- Microsoft Outlook
- Microsoft One Drive
- Skype for business
- Microsoft Teams
- Microsoft Publisher
- Microsoft Access etc.

MS-Word Shortcut Keys

Introduction:-

MS-Word is software which is developed by Microsoft Corporation with various versions. It is a part of MS-Office. This application is used to create professional quality documents with various templates and formatting tools. It has use in offices, schools and other educational institutes, homes and any kind of organization. Forms, circulars, worksheets, question papers, assignment submissions etc. Its file extension is .docx.

It has various components like Title bar, Ribbon, Sign in Button, Status bar, Maximize, Minimize and close buttons, Document Page, Quick Access Toolbar, Ribbon, Menu Bar, Sub Menu bar and Scroll Bar etc.

Every person needs shortcuts in life for work, so let's start with MS-Word with the help of these shortcut keys. So that work can be done very fast.

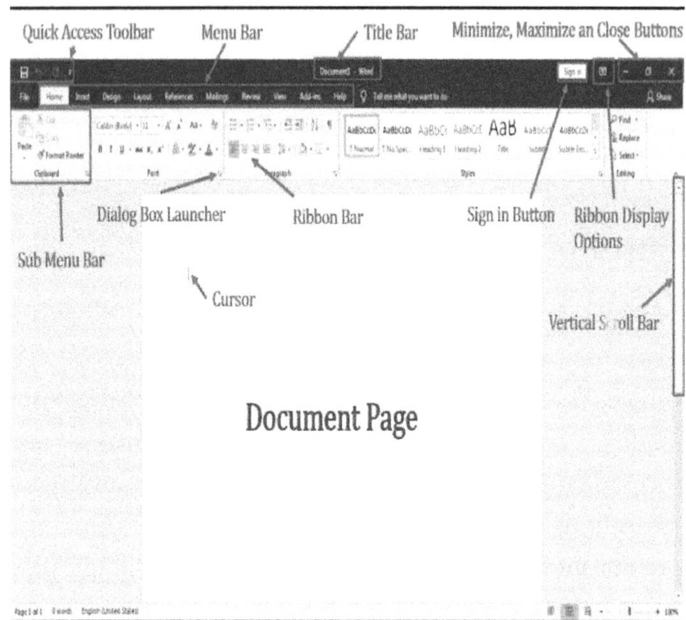

Shortcut Keys in MS-Word

Shortcut		Action
Ctrl+A	→	Selects all of the text in the document.
Ctrl+B	→	Makes the text bold.
Ctrl+C	→	Copy the text.
Ctrl+D	→	Open the Font dialog box to change the formatting of characters.
Ctrl+E	→	Aligns the line or selected text to the center of the page.

Ctrl+F	→	To find the text.
Ctrl+G	→	Move to a specific line in a document, spreadsheet, or text file. Display the GO TO dialog box.
Ctrl+H	→	Replace text in a word document.
Ctrl+I	→	Italics the selected text.
Ctrl+J	→	Aligns the selected text or line to justify the screen.
Ctrl+K	→	Inserts a hyperlink at the text cursor's current location.
Ctrl+L	→	Left align a paragraph or text.
Ctrl+M	→	Indents the paragraph or text.
Ctrl+N	→	Create a new document.
Ctrl+O	→	Open a document.
Ctrl+P	→	Opens the print preview window.
Ctrl+Q	→	Used to remove the paragraph's formatting.
Ctrl+R	→	Aligns the line or selected text to the right of the screen.
Ctrl+S	→	Saves the current document.
Ctrl+T	→	"Hangs" a paragraph to the next tab stop.
Ctrl+U	→	"underline." the text in document.

Ctrl+V	→	Pastes any text, image, video, or other object into the active document.
Ctrl+W	→	Close the document.
Ctrl+X	→	Cuts any text, picture, or other object that is selected.
Ctrl+Y	→	Redo any previously-undone action.
Ctrl+Z	→	Undo any change made in the document.
Ctrl+1	→	Apply single spacing to the paragraph.
Ctrl+2	→	Apply double spacing to the paragraph.
Ctrl+Shift+Right angle bracket(>)	→	To increase the font size of selected text.
Ctrl+Shift+Left angle bracket(<)	→	To decrease the font size of selected text.
Ctrl+Shift+8	→	To display non printing characters (but don't press 8 from numeric keypad).
Shift+Enter	→	Insert a line break.
Ctrl+Enter	→	Insert a page break.
Ctrl+Shift+K	→	To apply small caps formatting.
Ctrl+=(Equal)	→	Apply subscript formatting.
Ctrl+Shift+(plus) sign	→	Apply superscript formatting.
Ctrl+Alt+M	→	To insert a comment.

Alt+Shift+D	→	Insert a Date field.
Ctrl+Alt+L	→	Insert a Listnum field.
Alt+Shift+P	→	Insert a Page field.
Alt+Shift+T	→	Insert a Time field.
Alt+W,F	→	Opens Full Screen Reading View.
Ctrl+Alt+P	→	Opens Print Layout View.
Ctrl+Alt+O	→	Opens Outline View.
Ctrl+Alt+N	→	Opens Draft View.
F1	→	It displays the WORD help task pane.
F2	→	Moves the selected text or graphic and then press the enter key.
Shift + F3	→	It switches the selected text between upper case, lower case, and title case.
F4	→	It repeats the last command or action.
F5	→	It displays the Go To dialog.
F6	→	Switch between the document, task pane, status bar, and ribbon.
F7	→	It displays the Editor task pane to check spelling and grammar.

F8	→	It extends the selection, i.e. one word to one sentence.
F9	→	It updates the selected fields.
F10	→	It turns Key Tips on or off.
F11	→	It moves to next field.
F12	→	It displays the Save As dialog box.
Ctrl+Shift+Alt+S	→	Opens the Style Gallery.
Ctrl+Shift+N	→	Opens the Normal Style.
Ctrl+Alt+1	→	To Apply Heading 1 Style.
Ctrl+Alt+2	→	To Apply Heading 2 Style.
Ctrl+Alt+3	→	To Apply Heading 3 Style.
Ctrl+Shift+C	→	To copy the formatting of text.
Ctrl+Shift+V	→	To paste the formatting of text.
Ctrl+Shift+M	→	To Create a hanging indent.
Esc	→	To cancel the command
Ctrl+Alt+S	→	To Split the document window.
Ctrl+Alt+S or Alt+Shift+C	→	To remove the document window split.

Tab or Shift Tab	→	To move the focus to commands on the ribbon.
Alt+ Left arrow or Right arrow	→	To move between different tabs on ribbon.
Ctrl+F1	→	To Expand and Collapse the ribbon.
Shift+F10	→	To open the context menu.
Ctrl+ Left arrow key	→	To move the cursor one word left.
Ctrl+ Right arrow	→	To move the cursor one word right.
Ctrl+Alt+PageUp	→	Move the cursor to the top of the screen.
Ctrl+Alt+Page Down	→	Move the cursor to the bottom of the screen.
Ctrl+End	→	Move the cursor to the end of the document.
Ctrl+Home	→	Move the cursor to the beginning of the document.
Ctrl+Backspace	→	Delete one word to left direction.
Ctrl+Delete	→	Delete one word to right direction.
Alt+N,W	→	To insert a WordArt.
Alt+H	→	To open the Home tab.
Alt+N	→	To open the insert tab to insert tables, pictures and more commands in it.
Alt+P	→	To open the Page Layout tab.

Alt+R	→	To open the Review tab.
Alt+W	→	To open the View tab.
Alt+JT	→	To open the Design tab when table is drawn.
Alt+JL	→	To open the Layout tab when table is drawn.
Alt+M	→	To open the Mailings tab.
Alt+S	→	To open the Reference tab.
Alt+Shift+K	→	To preview the mail merge.
Alt+Shift+N	→	To preview the merge a document.
Alt+Shift+M	→	To print the merged document.
Alt+H+FP	→	To use format painter tool.
Alt+H+V	→	To use paste option.

MS-Excel Shortcut Keys

Introduction:-

MS-Excel is spreadsheet software which is developed by Microsoft Corporation with various versions. It is a part of MS-Office. It is a grid of cells arranged in rows and columns to organize data manipulations like arithmetic operations. It has use in offices, schools and other educational institutes, homes and any kind of organization. Rows are labeled as numbers (1, 2.....so on.) and columns are labeled as Alphabet (A, B....... so on.) and file extension is .xls/.xlsx .

Usually the **number of rows and columns** in a spreadsheet is 10, 48,576 **rows** and 16,384 **columns**.

The last column name in single spreadsheet is XFD.

It has various components like Title bar, Row bar, Column bar, Menu bar, Dialog box launcher, View buttons, Toolbar/Ribbon Formula bar, Sheet tabs, Active cell, Active Cell Reference (Name Box), Status bar, Scroll bars, Sheets and cells etc.

The first window of MS-Excel is given in figure here.

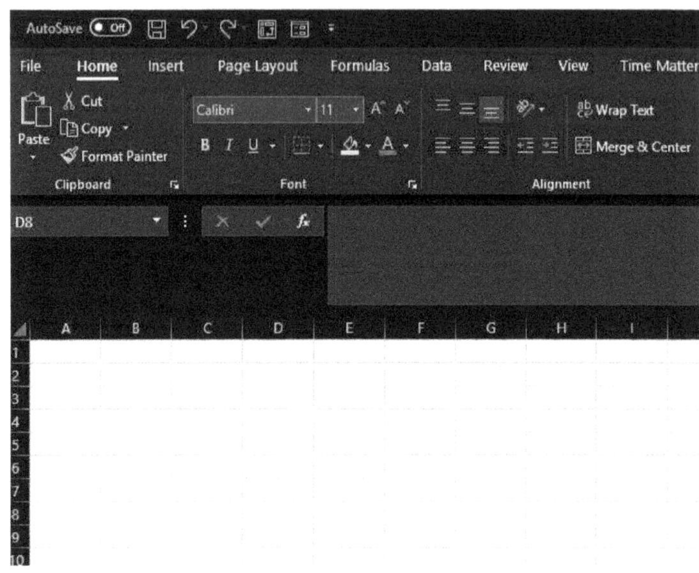

Shortcut Keys in MS-Excel

Shortcut		Action
Ctrl+A	→	Select the entire worksheet.
Ctrl+B	→	One or more cells is selected will bold or un-bold the selected cells.
Ctrl+C	→	When a cell(s) is highlighted copies its contents to the clipboard.
Ctrl+D	→	Overwrites a cell(s) with the contents of the cell above it in a column.

Ctrl+E	→	Displays or hides the outline symbols, Hides the selected rows, Hides the selected columns.
Ctrl+F	→	Opens the find box that allows you to search for characters, text, and phrases within a spreadsheet.
Ctrl+G	→	Opens the Go To window that allows you to focus a specific reference cell).
Ctrl+H	→	Opens the find and replace feature that allows you to find any text and replace it with any other text.
Ctrl+I	→	To Apply or remove italic formatting.
Ctrl+K	→	Inserts a hyperlink in the currently-active cell or location.
Ctrl+L	→	Opens the Create Table dialog box. If you're editing the contents of a cell.
Alt+M	→	Go to Formula tab.
Ctrl+N	→	Create a new workbook.
Ctrl+O	→	Open an existing workbook.
Ctrl+P	→	Opens the print preview window.
Ctrl+R	→	Fills the row cell to the right with the contents of the selected cell.

Ctrl+S	→	Save a workbook.
Ctrl+T	→	Opens the Create Table dialog box.
Ctrl+U	→	Add or remove underline to the contents of a cell, selected data, or selected cell range.
Ctrl+V	→	Paste anything stored in the clipboard into the currently-selected cell.
Ctrl+W	→	Closes the open workbook.
Ctrl+X	→	Cuts any cell, text, or other object that is selected in the spreadsheet.
Ctrl+Y	→	Redo any previously-undone action.
Ctrl+Z	→	Undo any change made in a spreadsheet.
F1	→	To open Help Menu.
F2	→	Go to Cell /Go to out of cell.
F3	→	Paste name in function.
F4	→	For Absolute reference.
F5	→	Go to
F6	→	Next Pane
F7	→	Spell check

F8	→	To Select anchor
F9	→	To Calculate various functions on values.
F10	→	To Activate menu bar.
F11	→	To Create chart sheet.
F12	→	Save as.
Ctrl+ SHIFT +	→	Select current region around active.
SHIFT + Arrow	→	Extend selection by one cell.
Ctrl + SHIFT + Arrow	→	Extend selection to last nonblank cell.
SHIFT + HOME	→	Extend selection to start of row.
Ctrl + SHIFT + HOME	→	Extend selection to top left.
Ctrl + SHIFT + End	→	Extend selection to last cell used.
Ctrl + Spacebar	→	Select the entire column of your active or selected cells.
SHIFT + Backspace	→	Only select active cell, when multiple cells are selected.
SHIFT + Page down	→	Extend selection down one screen.
SHIFT + Page up	→	Extend selection up one screen.
End	→	Turn End mode on or off.

SHIFT + Arrow	→	Extend selection to the last nonblank cell.
Alt + Enter	→	Start a new line in the same cell.
SHIFT + F2	→	Edit a cell comment.
SHIFT + Enter	→	Finish entry and move up in a selection.
SHIFT + Tab	→	Finish entry and move left in selection.
Ctrl + Delete	→	Delete text to the end of the line.
Ctrl + Enter	→	Fill the selected cell range with entry.
Tab	→	Finish entry and move right in selection.
ALT + '	→	Display the STYLE.
Ctrl + SHIFT + ~	→	General number format.
Ctrl + SHIFT + #	→	1-Jan-00 format.
Ctrl + SHIFT + $	→	Currency format.
Ctrl + SHIFT + _	→	Remove outline border.
Ctrl + SHIFT + ^	→	###E + 02 format.
Ctrl + 1	→	It brings Format Cells dialog box.
Ctrl + SHIFT +!	→	Value up to 2 decimals.

Ctrl + SHIFT + %	→	Percentage format.
Ctrl + SHIFT + &	→	Outline border.
Ctrl + SHIFT + @	→	1/1/1900 0:00 AM format.
Alt+H	→	To go to Home tab
Alt+N	→	To go to Insert tab in Ribbon.
Alt+P	→	To go to Page Layout tab in Ribbon.
Alt+M	→	To go to Formulas tab in Ribbon.
Alt+A	→	To go to Data tab.
Alt+R	→	To go to Review tab.
Alt+W	→	To go to View tab.
Ctrl+9	→	To hide the selected row.
Ctrl+0	→	To hide the selected column.

MS-PowerPoint Shortcut Keys

Introduction:-

MS-PowerPoint is Presentation software which is developed by Microsoft Corporation with various versions. It is a part of MS-Office. It is a collection of slides (page) in sequential manner to present ideas to audience. It has use in offices, schools, business meetings and other educational institutes, homes and any kind of organization. Various slide layouts are available in it to arrange the data on slide. Its file extension is .pptx and the default name of presentation is Presentation 1, 2, 3........so on.

It has various components like File Menu and Backstage View, Quick Access Toolbar (QAT), Ribbon, Slides Pane, Slide Area, Task Pane, Status Bar, Notes Pane etc.

The first window of MS-PowerPoint is given here:-

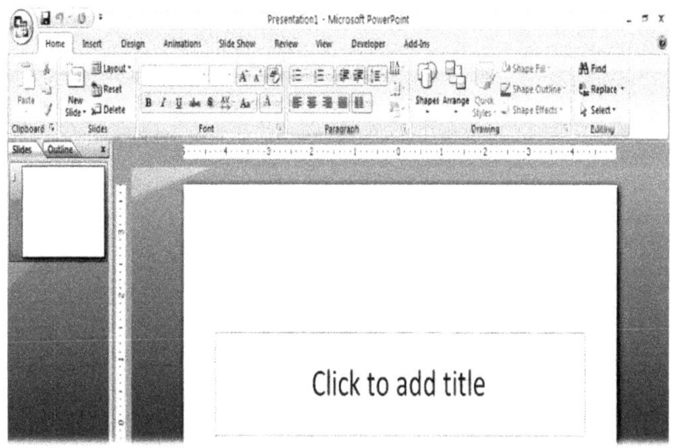

Shortcut Keys in MS-Power Point

Shortcut		Action
Ctrl+A	→	Select all the objects on an active slide.
Ctrl+B	→	To bold the text in slide.
Ctrl + C or Ctrl + Insert	→	Copy the selected text, object, or selected slide.
Ctrl+D	→	Inserts a duplicate of the selected slide.
Ctrl+E	→	Center text within a box.
Ctrl+F	→	Opens the find window.
Ctrl+G	→	Group shapes easily.
Ctrl+H	→	Hides the cursor or any activated tools, like the pen

		or highlighter.
Ctrl+I	→	Apply or remove italic formatting.
Ctrl+J	→	Aligns, or distributes, the text evenly across the slide.
Ctrl+K	→	Insert hyperlink
Ctrl+L	→	Left align text within a box.
Ctrl+M	→	Inserts a blank slide after the currently-selected one.
Ctrl+N	→	Create a new presentation document.
Ctrl+O	→	Open an existing presentation document.
Ctrl+P	→	Annotate using a Pen tool while playing the slideshow.
Ctrl+Q	→	Save and close a presentation.
Ctrl+R	→	Right align text within a box.
Ctrl+S	→	Save a presentation.
Ctrl+T	→	Open the Font window, to adjust font size, style, and type.
Ctrl+U	→	Add or remove underline to the contents of a slide.
Ctrl + V or Shift + Insert	→	Paste the selected text, object, or slide.

Ctrl+W	→	Closes the current presentation.
Ctrl+X	→	Cut the selected text, object, or slide.
Ctrl+Y	→	Redo an action.
Ctrl+Z	→	Undo any edit made to a presentation.
Tab	→	Select or move on to the next object on a slide.
Shift + Tab	→	Select or move to a previous object on a slide.
Home	→	Go back to the beginning of the slide.
End	→	Go to the end of the slide.
PgDn	→	Go to the next slide.
PgUp	→	Go the previous slide.
Ctrl + Up / Down Arrow	→	Move a slide up or down in the presentation.
Ctrl + Shift + Up / Down Arrow	→	Move a slide to the beginning or end of your presentation document.
F5	→	Play the presentation from the start.
Shift + F5	→	Play the presentation from the current slide.
N or Page Down	→	Move to the next slide while playing the slideshow.
P or Page Up	→	Return to the previous slide while playing the

		slideshow.
B	→	Change the screen to black during a slideshow (Press B again to return to the slideshow).
Esc	→	End the slideshow.

Adobe Photoshop Shortcut Keys

Introduction:-

Adobe Photoshop software is the industry standard in digital imaging and is used worldwide for design, photography, video editing and more. It is published by Adobe Inc. It was originally created in 1988 by Thomas and John Knoll, who sold the distribution license to Adobe Systems Incorporated in 1988.Its latest release is version (22.4.1) in May 2021.It is available in 26 languages. The extension of Adobe Photoshop Document is .PSD.

The first window of Adobe Photoshop is given here.

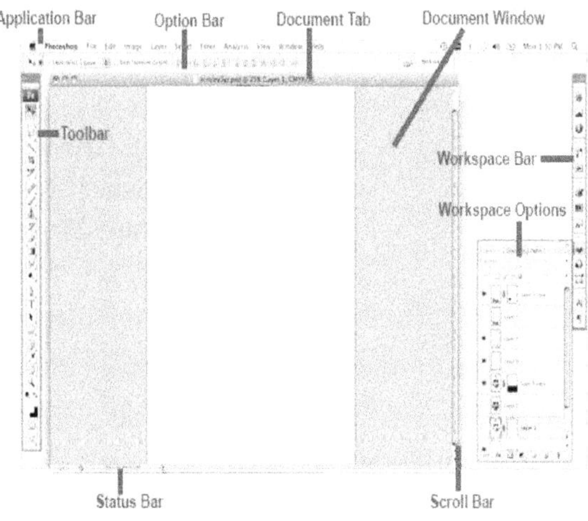

Shortcut		Action
Shift+A	→	Path Selection, Direct Selection.
Shift+B	→	Brush, Pencil, Color Replacement, Mixer Brush.
Shift+C	→	Crop and Slice Tools.
Shift+D	→	Default colors.
Shift+E	→	Eraser tools.
Shift+G	→	Gradient, Paint Bucket, 3D Material Drop.
Shift+H	→	Hand tool
Shift+I	→	Eyedropper, 3D Material Eyedropper, Color Sampler, Ruler, Note, Count.
Shift+J	→	Spot Healing Brush, Healing Brush, Patch, Content-Aware Move, Red Eye.
Shift+L	→	Lasso tools
Shift+M	→	Marquee tools
Shift+O	→	Dodge, Burn, Sponge.

Shift+P	→	Pen tools
Shift+Q	→	Quick Mask Mode
Shift+R	→	Rotate View
Shift+S	→	Clone Stamp, Pattern Stamp.
Shift+T	→	Type tools
Shift+U	→	Rectangle, Rounded Rectangle, Ellipse, Polygon, Line, Custom Shape.
Shift +Vor Shift + Insert	→	Move and Art board tools.
Shift+W	→	Quick Selection, Magic Wand.
Shift+X	→	Switch Foreground and Background colors.
Shift+Y	→	History Brush, Art History Brush.
Shift+Z	→	Zoom
Ctrl+T	→	Free Transform
[	→	Decrease brush size
]	→	Increase brush size
{	→	Decrease brush hardness
}	→	Increase brush hardness

D	→	Default Foreground/Background colors.
X	→	Switch Foreground/Background colors.
Alt-click layer	→	Fit layer(s) to screen
Ctrl+J	→	New layer via copy
Shift + Ctrl + J	→	New layer via cut
Any selection tool + Shift-drag	→	Add to a selection
Alt-click brush or swatch	→	Delete brush or swatch
Ctrl-click	→	Toggle auto-select with the move tool.
Ctrl + Alt + P	→	Close all open documents other than the current document.
Escape	→	Cancel any modal dialog window (including the Start Workspace).
Enter	→	Select the first edit field of the toolbar.
Tab	→	Navigate between fields.
Tab + Shift	→	Navigate between fields in the opposite direction.
Alt	→	Change Cancel to Reset.
F1	→	Start Help

F2	→	Cut
F3	→	Copy
F4	→	Paste
F5	→	Show/Hide Brush panel.
F6	→	Show/Hide Color panel.
F7	→	Show/Hide Layers panel.
F8	→	Show/Hide Info panel.
F9	→	Show/Hide Actions panel.
F12	→	Revert
Shift + F5	→	Fill
Shift + F6	→	Feather Selection
Shift + F7	→	Inverse Selection
Control + Shift + T	→	Repeat last duplicate and move.
Shift-Click	→	Select/deselect multiple contiguous layers.
Control-click	→	Select/deselect multiple discontiguous layers.
Alt + . (period)	→	Select top layer

Alt + ,(comma)	→	Select bottom layer
Ctrl + G	→	Group layers.
Ctrl+Shift+G	→	UnGroup layers.
Ctrl + Alt + G	→	Create/release clipping mask.
Ctrl + Alt + A	→	Select all layers.
Ctrl+Shift+E	→	Merge visible layers.
Ctrl + Shift + Alt + E	→	Merge a copy of all visible layers into target layer.
Alt-click the eye icon	→	Show/hide all other currently visible layers.
/(forward slash)	→	Toggle lock transparency for target layer, or last applied lock.
Double-click the filter effect	→	Edit filter settings.

Corel PaintShop Pro Shortcut Keys

Introduction:-

Corel PaintShop Pro (PSP) is a raster and vector graphics editor for Microsoft Windows. It was originally published by Jasc Software, Inc. In October 2004, Corel purchased Jasc Software and the distribution rights to Paint Shop Pro. Its file extension is .PSP.

The first window of Corel PaintShop Pro is given here.

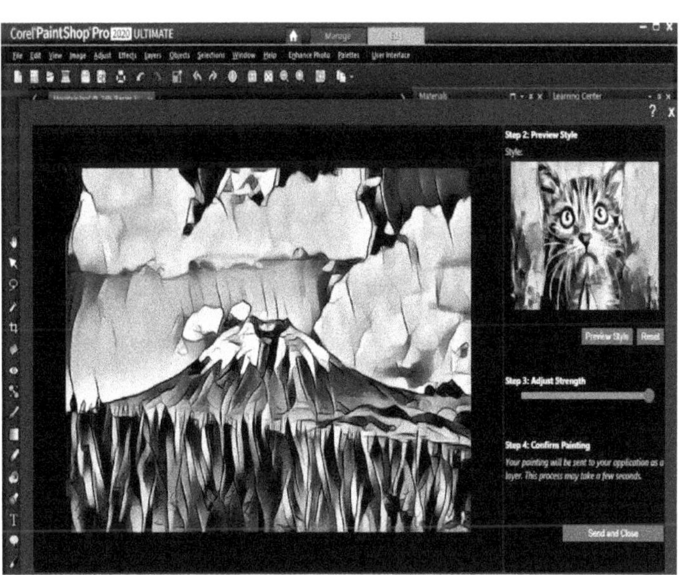

Shortcut Keys in Corel PaintShop Pro

Shortcut		Action
Ctrl + N	→	New
Ctrl + O	→	Open
Ctrl + B	→	Browse
Ctrl + S	→	Save
F12	→	Save As
Ctrl + F12	→	Save Copy As
Ctrl + Del	→	Delete
Ctrl + P	→	Print
Ctrl + Z	→	Undo
Shift + Ctrl + Z	→	Undo History
Ctrl + X	→	Cut
Ctrl + C	→	Copy
Shift + Ctrl + C	→	Copy Merged.
Ctrl + V	→	Paste As New Image.
Ctrl + L	→	Paste as New Layer.
Ctrl + E	→	Paste as New Selection.
Shift + Ctrl + E	→	Paste as a Transparent Selection.
Shift + Ctrl + L	→	Paste into Selection.
Del	→	Clear.
Shift + A	→	Full Screen Edit.
Shift+ Ctrl + A	→	Full Screen Preview.

Ctrl + Alt + N	→	Normal Viewing
Shift + I	→	Image. Information
Ctrl + Alt + G	→	Grid.
Ctrl + I	→	Flip
Ctrl + M	→	Mirror
Ctrl + R	→	Rotate
Shift + R	→	Crop to Selection.
Shift + S	→	Resize
Shift + Y	→	Hide All
Shift + K	→	Invert
Ctrl + K	→	Edit
Ctrl + Alt + V	→	View Mask
Shift + B	→	Brightness/Contrast.
Shift + G	→	Gamma Correction.
Shift + M	→	Highlight/ Midtone /Shadow.
Shift + H	→	Hue/Saturation/Luminescence.
Shift + U	→	Red/Green/Blue.
Shift + L	→	Colorize
Shift + Z	→	Posterize
Shift + P	→	Edit Palette
Shift + D	→	Load Palette
g	→	Zoom
d	→	Deform
r	→	Crop
v	→	Mover
s	→	Selection

a	→	Freehand Selection.
m	→	Magic Wand Selection.
y	→	Eye Dropper
b	→	Paintbrush
n	→	Clone
, {comma}	→	Color Replacer
z	→	Retouch
e	→	Eraser
. {period}	→	Picture Tube
u	→	Airbrush
f	→	Flood Fill
x	→	Text
i	→	Line
/ {slash}	→	Shape
Shift + W	→	New Window
Shift + D	→	Duplicate
Ctrl + W	→	Fit to Window
Ctrl + A	→	Select All
Ctrl + D	→	Select None
Shift + Ctrl + S	→	From Mask
Shift + Ctrl + M	→	Hide Selection Marquee.
Shift + Ctrl + I	→	Invert
Shift + Ctrl + P	→	Promote Layer
Ctrl + F	→	Float
Shift + Ctrl + F	→	Defloat

Adobe Page Maker Shortcut Keys

Introduction:-

Adobe PageMaker 7.0 was the final version made available. It was released 9 July 2001. It's made for designing things for print, from flyers and posters to reports, and will export creations as PDF files too. Its file extensions are .pmd/.pmt/.pm3/.pm4/.pm5/.pm6 etc. The first window of Adobe Page Maker is given here.

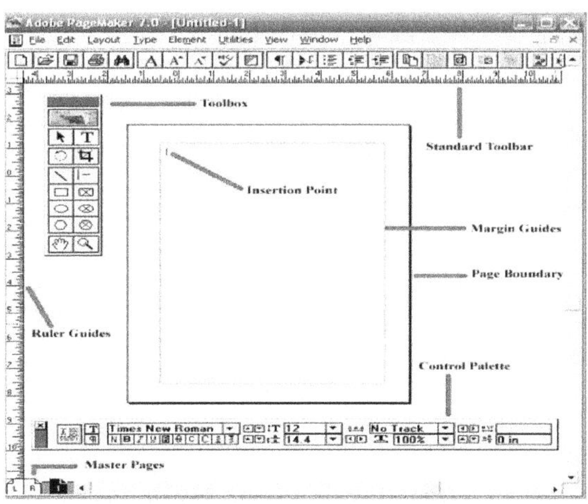

Shortcut Keys in Adobe Page Maker

Shortcut		Action
Ctrl+ N	→	New
Ctrl+ O	→	Open
Ctrl+ S	→	Save
Ctrl+ W	→	Close
Ctrl+Shift+S	→	Save As
Ctrl + D	→	Place
Ctrl+Shift+D	→	Link Manager
Ctrl+Shift+P	→	Document Setup
Ctrl+P	→	Print
Ctrl + K	→	General
Ctrl+ Q	→	Quite
Ctrl + A	→	Select All
Ctrl + C	→	Copy
Ctrl + V	→	Paste
Ctrl + Z	→	Undo
Ctrl + E	→	Edit story
Alt+Ctrl+G	→	Go to
Ctrl+Shift+>	→	Increase Font Size.
Ctrl+Alt+>	→	Standard Size
Ctrl+Shift+>	→	Increase Font size.
Ctrl+Shift+<	→	Decrease Font Size.
Ctrl+Shift+B	→	Bold
Ctrl+Shift+I	→	Italic

Ctrl+Shift+U	→	Underline
Ctrl+Shift+C	→	Center Align
Ctrl+Shift+J	→	Justify
Ctrl+Shift+L	→	Left Align
Ctrl+Shift+R	→	Right Align
Ctrl+Shift+/	→	Strikethrough
Ctrl+Shift+V	→	Reverse
Ctrl+T	→	Character Specs
Ctrl+Shift+K	→	All Caps
Ctrl+	→	Subscript
Ctrl+Shift+	→	Superscript
Ctrl+M	→	Paragraph
Ctrl+I	→	Indents/Tabs
Ctrl+3	→	Define Style
Ctrl+U	→	File & Stroke
Alt + Ctrl+F	→	Frame Option
Ctrl+}	→	Bring To Front
Ctrl+{	→	Send To Back
Ctrl+Shift+}	→	Send Backward
Ctrl+Shift+{	→	Bring Forward
Ctrl+Shift+E	→	Align Object
Ctrl+Alt+E	→	Text Wrap
Ctrl+G	→	Group
Ctrl+Shift+G	→	Ungroup
Ctrl+6	→	Mask

Ctrl+Shift+6	→	Unmask
Ctrl+L	→	Lock Position
Alt+Ctrl+L	→	Unlock
Ctrl+F	→	Find
Ctrl+H	→	Change
Ctrl+L	→	Spelling
Ctrl+Y	→	Index Entry
Ctrl+ +	→	Zoom In
Ctrl+-	→	Zoom Out
Ctrl+1	→	Actual Size
Ctrl+2	→	200 Percent
Ctrl+4	→	400 Percent
Ctrl +5	→	500 Percent
Ctrl+7	→	75 percent
Ctrl+0	→	Fit In Window
Ctrl+R	→	Show/Hide Ruler
Ctrl+J	→	Show/hide Color

CorelDRAW Shortcut Keys

Introduction:-

CorelDRAW is a graphics editor developed and marketed by Corel Corporation. It is also the name of the Corel graphics suite, which includes the bitmap-image editor Corel Photo-Paint as well as other graphics-related programs. The latest version is marketed as CorelDRAW Graphics Suite 2021, and was released in March, 2021. CorelDRAW is designed to edit two-dimensional images such as logos and posters and it is available for Windows and Mac OS. Its file extension is .cdr file. The first window of CorelDRAW is given here.

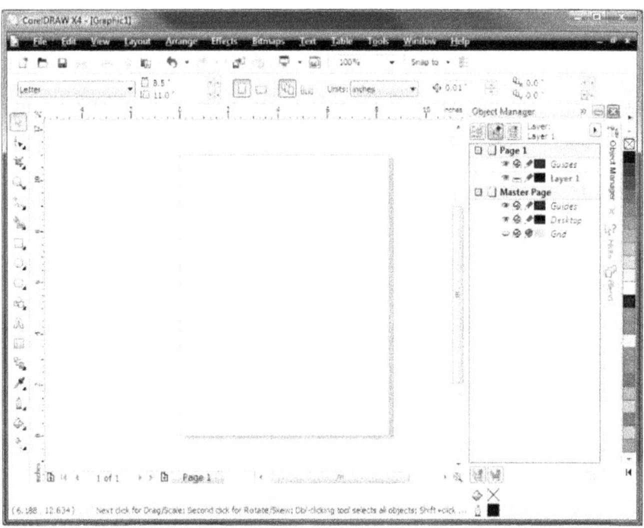

Shortcut Keys in CorelDRAW

Shortcut		Action
Ctrl+A	→	Select all the objects
Ctrl+B	→	Bold the text
Ctrl+C	→	Copy the Objects
Ctrl+D	→	Create Duplicate
Ctrl+E	→	Export Work
Ctrl+F	→	Ctrl + F Search Feature.
Ctrl+G	→	Group the objects
Ctrl+I	→	Import data
Ctrl+J	→	Option
Ctrl+K	→	Break Apart
Ctrl+N	→	New
Ctrl+O	→	Open
Ctrl+P	→	Print
Ctrl+Q	→	Curve
Ctrl+S	→	Save
Ctrl+T	→	Format text
Ctrl+U	→	Ungroup
Ctrl+V	→	Paste
Ctrl+X	→	Cut
Ctrl+Z	→	Undo
F1	→	Corel draw help
F2	→	Zoom tool
F3	→	To reduce size
F4	→	Fit in window

F5	→	Free hand tool
F6	→	Rectangle Tool
F7	→	Ellipse tool circle
F8	→	Text tool
F9	→	Full screen
F10	→	Shape tool
F11	→	Fountain fill
B	→	Aligns selected objects to the bottom.
E	→	Horizontally aligns the centers of the selected objects.
C	→	Vertically aligns the centers of the selected objects.
L	→	Aligns selected objects to the left.
R	→	Aligns selected objects to the right.
Alt+F12	→	Aligns text to the baseline
T	→	Aligns selected objects to the top.
I	→	Draws curves and applies Preset, Brush, and Spray, Calligraphic or Pressure Sensitive effect.
Ctrl+PgDn	→	Back One
Ctrl+K	→	Breaks apart the selected object.
Ctrl+B	→	Brightness/Contrast/Intensity.
Ctrl+Enter	→	Brings up the Property Bar and gives focus to the first visible item that can be

		tabbed.
P	→	Aligns the centers of the selected objects to page.
Ctrl+T	→	Character Formatting.
Ctrl+Shift+	→	Color Balance
Ctrl+L	→	Combines the selected objects.
Ctrl+F9	→	Opens the Contour Docker Window.
Ctrl+F8	→	Converts artistic text to paragraph text or vice versa.
Ctrl+Shift+Q	→	Converts an outline to an object..
Ctrl+Q	→	Converts the selected object to a curve.
Ctrl+C	→	Copies the selection and places it on the Clipboard.
Ctrl+Insert	→	Copies the selection and places it on the Clipboard.
Ctrl+X	→	Cuts the selection and places it on the Clipboard.
Delete	→	Deletes the selected object(s).
Shift+B	→	Distributes selected objects to the bottom.
Shift+E	→	Horizontally Distributes the centers of the selected objects.
Shift+C	→	Vertically Distributes the centers of the selected objects.
Shift+L	→	Distributes selected objects to the left.

Shift+R	→	Distributes selected objects to the right.
Shift+P	→	Horizontally Distributes the space between the selected objects.
Shift+A	→	Vertically Distributes the space between the selected objects.
Shift+T	→	Distributes selected objects to the top.
Ctrl+D	→	Duplicates the selected object(s) and offsets by a specified amount.
Alt+Shift+D	→	Shows or hides the Dynamic Guides (toggle).
Ctrl+Shift+T	→	Opens the Edit Text dialog box.
F7	→	Draws ellipses and circles; double-clicking the tool opens the Toolbox tab of the Option.
Ctrl+F7	→	Opens the Envelope Docker Window.
X	→	Erases part of a graphic or splits an object into two closed paths.
Alt+F4	→	Exits CorelDRAW and prompts to save the active drawing.
Ctrl+E	→	Exports text or objects to another format.
Ctrl+NUMPAD2	→	Decreases font size to previous point size.
Ctrl+NUMPAD8	→	Increases font size to next

		point size.
Ctrl+NUMPAD6	→	Increase font size to next setting in Font Size List.
Ctrl+NUMPAD4	→	Decrease font size to previous setting available in the Font Size List.
Ctrl+PgUp	→	Forward One
F11	→	Applies fountain fills to objects.
F5	→	Draws lines and curves in freehand mode.
F9	→	Displays a full-screen preview of the drawing.
D	→	Draws a group of rectangles; double-clicking opens the Toolbox tab of the Options dial.
Ctrl+F5	→	Opens the Graphic and Text Styles Docker Window.
Ctrl+G	→	Groups the selected objects.
H	→	Hand Tool
Ctrl+,	→	Changes the text to horizontal direction.
Ctrl+Shift+U	→	Hue/Saturation/Lightness.
Ctrl+I	→	Imports text or objects.
Ctrl+F11	→	Opens the Insert Character Docker Window.
G	→	Adds a fill to object(s); clicking and dragging on object(s) applies a fountain fill.
Alt+F3	→	Opens the Lens Docker Window.

Alt+F2	→	Contains functions for assigning attributes to linear dimension lines.
Alt+F11	→	Macro Editor...
M	→	Converts an object to a Mesh Fill object.
Ctrl+DnArrow	→	Nudges the object downward by the Micro Nudge factor.
Ctrl+LeftArrow	→	Nudges the object to the left by the Micro Nudge factor.
Ctrl+RightArrow	→	Nudges the object to the right by the Micro Nudge factor.
Ctrl+UpArrow	→	Nudges the object upward by the Micro Nudge factor.
N	→	Brings up the Navigator window allowing you to navigate to any object in the document.
Ctrl+N	→	Creates a new drawing.
PgDn	→	Goes to the next page.
DnArrow	→	Nudges the object downward.
LeftArrow	→	Nudges the object to the left.
RightArrow	→	Nudges the object to the right.
UpArrow	→	Nudges the object up.
Ctrl+O	→	Opens an existing drawing.
Ctrl+J	→	Opens the dialog for setting CorelDraw options.

Shift+F12	→	Opens the Outline Color dialog box.
F12	→	Opens the Outline Pen dialog box.
Alt+DnArrow	→	Pan Down
Alt+LeftArrow	→	Pan Left
Alt+RightArrow	→	Pan Right
Alt+UpArrow	→	Pan Up
Ctrl+V	→	Pastes the Clipboard contents into the drawing.
Shift+Insert	→	Pastes the Clipboard contents into the drawing.
Ctrl+1	→	Places selected object(s) into a PowerClip container object.
Y	→	Draws polygons
Alt+F7	→	Opens the Position Docker Window.
PgUp	→	Goes to the previous page.
Ctrl+P	→	Prints the active drawing.
Alt+Enter	→	Allows the properties of an object to be viewed and edited.
Ctrl+Shift+R	→	Record Temporary Macro.
F6	→	Draws rectangles; double-clicking the tool creates a page frame.
Ctrl+Shift+Z	→	Reverses the last Undo operation.
Ctrl+W	→	Redraws the drawing window.
Ctrl+R	→	Repeats the last operation.

Alt+F8	→	Opens the Rotate Docker Window.
Ctrl+Shift+P	→	Run Temporary Macro.
Ctrl+S	→	Saves the active drawing.
Alt+F9 Window	→	Opens the Scale Docker.
Ctrl+A	→	Select all object of the active page.
F10	→	Edits the nodes of an object; double-clicking the tool selects all nodes on the selected.
Alt+F10 Window	→	Opens the Size Docker.
Ctrl+Y	→	Snaps objects to the grid (toggle).
Alt+Z	→	Snaps objects to other objects (toggle).
Ctrl+F12	→	Opens the Spell Checker; checks the spelling of the selected text.
A	→	Draws spirals; double-clicking opens the Toolbox tab of the Options dialog.
Ctrl+Shift+D	→	Shows Step and Repeat docker.
Ctrl+Shift+O	→	Stop Recording.
Shift+DnArrow	→	Nudges the object downward by the Super Nudge factor.
Shift+LeftArrow	→	Nudges the object to the left

		by the Super Nudge factor.
Shift+RightArrow	→	Nudges the object to the right by the Super Nudge factor.
Shift+UpArrow	→	Nudges the object upward by the Super Nudge factor.
Ctrl+F3	→	Symbol Manager Docker.
F8	→	Adds text; click on the page to add Artistic Text; click and drag to add Paragraph Text.
Shift+PgDn	→	To Back Of Layer.
Ctrl+End	→	To Back Of Page.
Shift+PgUp	→	To Front Of Layer.
Ctrl+Home	→	To Front Of Page.
Ctrl+Space	→	Toggles between the current tool and the Pick tool.
Shift+F9	→	Toggles between the last two used view qualities.
Ctrl+Z	→	Reverses the last operation.
Alt+Backspace	→	Reverses the last operation.
Ctrl+U	→	Ungroup the selected objects or group of objects.
Shift+F11	→	Applies uniform color fills to objects.
Ctrl+M	→	Show/Hide Bullet.
Ctrl+.	→	Changes the text to vertical.
Ctrl+F2	→	Opens the View Manager Docker Window.
Shift+F1	→	What's This? Help.
Z	→	Zoom Tool
F3	→	Zoom Out

F4 → Zoom To All Objects.

Shift+F4 → Zoom To Page.

Shift+F2 → Zoom To Select.

Google Chrome Shortcut Keys

Action		Shortcut
Search with your default search engine.	→	Type a search term + Enter.
Search using a different search engine.	→	Type a search engine name and press Tab.
Add www. and .com to a site name, and open it in the current tab.	→	Type a site name + Ctrl + Enter.
Open a new tab and perform a Google search	→	Type a search term + Alt + Enter.
Jump to the address bar.	→	Ctrl + l or Alt + D or F6.
Search from anywhere on the page.	→	Ctrl + K or Ctrl + E.
Remove predictions from your address bar.	→	Down arrow to highlight + Shift + Delete.
Move cursor to the address bar.	→	Control + F5.

Address Bar Shortcut Keys

Action		Shortcut
Show or hide the Bookmarks bar.	→	Ctrl + Shift + B
Open the Bookmarks Manager.	→	Ctrl + Shift + O

Open the History page in a new tab.	→ **Ctrl + H**
Open the Downloads page in a new tab.	→ **Ctrl + J**
Open the Chrome Task Manager.	→ **Shift + Esc**
Set focus on the first item in the Chrome toolbar.	→ **Shift + Alt + T**
Set focus on the rightmost item in the Chrome toolbar.	→ **F10**
Switch focus to unfocused dialog (if showing) and all toolbars.	→ **F6**
Open the Find Bar to search the current page.	→ **Ctrl + F or F3**
Jump to the next match to your Find Bar search.	→ **Ctrl + G**
Jump to the previous match to your Find Bar search.	→ **Ctrl + Shift + G**
Open Developer Tools.	→ **Ctrl + Shift + J or F12.**
Open the Clear Browsing Data options.	→ **Ctrl + Shift + Delete.**
Open the Chrome Help Center in a new tab.	→ **F1**
Log in a different user or browse as a Guest.	→ **Ctrl + Shift + M.**
Open a feedback form	→ **Alt + Shift + I.**
Turn on caret browsing	→ **F7**

Chrome Features Shortcut Keys

Action	Shortcut
Open a new window in	→ **Ctrl + Shift + N.**

Incognito mode.

Open a new tab, and jump to it.	→	Ctrl + T
Reopen previously closed tabs in the order they were closed.	→	Ctrl + Shift + T
Jump to the next open tab.	→	Ctrl + Tab or Ctrl + PgDn.
Jump to the previous open tab.	→	Ctrl + Shift + Tab or Ctrl + PgUp.
Jump to a specific tab.	→	Ctrl +1 through Ctrl + 8.
Jump to the rightmost tab.	→	Ctrl + 9
Open your home page in the current tab.	→	Alt + Home
Open the previous page from your browsing history in the current tab.	→	Alt + Left arrow
Open the next page from your browsing history in the current tab.	→	Alt + Right arrow
Close the current tab.	→	Ctrl + W or Ctrl + F4.
Close the current window.	→	Ctrl + Shift + W or Alt + F4.
Minimize the current window.	→	Alt + Space then N.
Maximize the current window.	→	Alt + Space then X.
Quit Google Chrome.	→	Alt + F then X.

Action		Shortcut
Open options to print the current page.	→	Ctrl + P
Open options to save the current page.	→	Ctrl + S
Reload the current page.	→	F5 or Ctrl + R.
Reload the current page, ignoring cached content.	→	Shift + F5 or Ctrl + Shift + R.
Stop the page loading	→	Esc
Browse clickable items moving forward.	→	Tab.
Browse clickable items moving backward.	→	Shift + Tab.
Open a file from your computer in Chrome.	→	Ctrl + O + Select a file.
Display non-editable HTML source code for the current page.	→	Ctrl + U
Save your current webpage as a bookmark.	→	Ctrl + D
Save all open tabs as bookmarks in a new folder.	→	Ctrl + Shift + D.
Turn full-screen mode on or off.	→	F11
Make everything on the page bigger.	→	Ctrl and +
Make everything on the page smaller.	→	Ctrl and -
Return everything on the page	→	Ctrl + 0

to default size.		
Scroll down a webpage, a screen at a time.	→	**Space or PgDn.**
Scroll up a webpage, a screen at a time.	→	**Shift + Space or PgUp.**
Go to the top of the page.	→	**Home**
Go to the bottom of the page.	→	**End**
Scroll horizontally on the page.	→	**Shift + Scroll your mouse wheel.**
Move your cursor to the beginning of the previous word in a text field.	→	**Ctrl + Left arrow.**
Move your cursor to the next word.	→	**Ctrl + Right arrow.**
Delete the previous word in a text field.	→	**Ctrl + Backspace.**
Open the Home page in the current tab.	→	**Alt + Home.**
Reset page zoom level.	→	**Ctrl + 0**

Mouse Shortcut Keys in Chrome

Action		Shortcut
Open a link in a current tab (mouse only).	→	**Drag a link to a tab.**
Open a link in new background tab.	→	**Ctrl + Click a link.**
Open a link, and jump to it.	→	**Ctrl + Shift +Click a link.**
Open a link, and jump to it	→	**Drag a link to a**

(mouse only).		**blank area of the tab strip.**
Open a link in a new window.	→	**Shift + Click a link.**
Open a tab in a new window (mouse only).	→	**Drag the tab out of the tab strip.**
Move a tab to a current window (mouse only)	→	**Drag the tab into an existing window.**
Return a tab to its original position.	→	**Press Esc while dragging.**
Save the current webpage as a bookmark.	→	**Drag the web address to the Bookmarks Bar.**
Scroll horizontally on the page.	→	**Shift + Scroll your mouse wheel.**
Download the target of a link.	→	**Alt + Click a link.**
Display your browsing history.	→	**Right-click Back ← or click & hold Back ← Right-click Next → or click & hold Next →.**
Switch between maximized and windowed modes.	→	**Double-click a blank area of the tab strip.**
Make everything on the page bigger.	→	**Ctrl + Scroll your mouse wheel up.**
Make everything on the page smaller.	→	**Ctrl + Scroll your mouse wheel down.**

Windows Generation (Year wise)

- MS-DOS – Microsoft Disk Operating System (1981)
- Windows 1.0 – 2.0 (1985-1992)
- Windows 3.0 – 3.1 (1990-1994)
- Windows NT 3.1 – 4.0 (1993 – 1996)
- Windows 95 (August 1995)
- Windows 98 (June 1998)
- Windows 2000 (February 2000)
- Windows ME–Millennium Edition(September 2000)
- Windows Mobile (April 2000)
- Windows XP (October 2001)
- Windows Server (March 2003)
- Windows Vista (November 2006)
- Windows CE (November 2006)
- Windows Home Server (January 2007)
- Windows 7 (October, 2009)
- Windows Phone (November 2010)
- Windows 8(August, 2012)

- Windows 10(July, 2015)
- Windows 11(June, 2021)

Tips to Improve Performance in Windows 10

1. Restart device
2. Disable startup apps
3. Disable prelaunch apps on startup
4. Disable background apps
5. Uninstall non-essential apps
6. Install quality apps only
7. Clean up hard drive space
8. Use drive defragmentation
9. Configure Ready Boost
10. Perform malware scan
11. Install latest update
12. Switch to high performance power plan
13. Disable system visual effects
14. Disable search indexing
15. Increase page file size
16. Restore previous working state
17. Repair Windows setup files
18. Reset device to factory defaults
19. Upgrade to faster drive
20. Upgrade system memory

Hidden Features of Windows 10

Microsoft Windows isn't any one thing; it's an interwoven patchwork of tools built atop features that trace back to the beginning of the time-tested operating system.

With such a complex piece of software, it makes sense that there are little tricks and UI flourishes most people don't even know about. Maybe you haven't poked around Windows 10 too much after coming over from Windows 7, or perhaps you recently made the switch from a Mac. Well, it's time to understand all the secret Windows 10 has to offer.

We have compiled a list of useful tips that will help you get more out of your Windows 10 experience. Or, at least, teach you some things you may not have known about. Some have been available in Windows for a number of generations, while others are native to Windows 10.

Microsoft's most recent update for the OS arrived in May, introducing a bunch of new features with Windows 10 version 2004. So there are plenty of ways to make the most of a constantly evolving Windows experience.

1. Secret Start Menu

If you're a fan of that old-school (i.e. non-tiled) Start menu experience, you can still (sort of) have it. Right-click on the Windows icon in the bottom-left corner to prompt a textual jump menu with a number of familiar destinations, including Apps and Features, Search, and Run. All these options are available through the standard menu interface, but you will be able to access them quicker here.

Similarly, there's a lot you can do with the Windows 10 taskbar. Right-click on the taskbar for a handy menu that will allow you to quickly access a number of presets for the toolbars, Cortana, and window schemes.

Want to personalize those Start menu tiles? Right-click on them to prompt a pop-up menu. This menu will give you various options, like the ability to un-pin from the Start menu, resize the windows, or turn that live tile off.

2. Show Desktop Button

Dating back to Windows 7, the Show Desktop button is a handy little feature. On the bottom-right corner of the desktop is a secret button. Don't see it? Look all the way to the bottom and right, beyond the date and time. There you'll find a small little sliver of an invisible button. Click it to minimize all your open windows at once.

There's also the option to have windows minimize when you hover over this button versus clicking. Select your preference in Settings > Personalization > Taskbar, then flip the switch under "Use peek to preview the desktop."

3. Enhanced Windows Search

If searches are taking too long in Windows, you can narrow things down a bit thanks to the May 2020 Update. Under Settings > Search > Searching Windows set search to Classic, which only applies to Libraries and Desktop, or choose Enhanced indexing to search the whole computer. A new algorithm also helps Windows adjust when it's working, using less resource while gaming or when disk usage is over 80 percent.

4. Rotate Your Screen

If you use multiple displays, this feature allows you to orient a particular monitor to fit your needs. The quickest way to do this is to simultaneously press and hold Ctrl + Alt together, and then use a directional arrow to flip the screen. The right and left arrows turn the screen 90 degrees, while the down arrow will flip it upside down. Use the up arrow to bring the screen back to its normal position.

These key commands only work with certain computers, so if you can't get them to work, you can go through Settings > System > Display, or right-click on the desktop and choose Display Settings to get there faster. Choose an option from the Display Orientation drop-down menu to turn your page around in all sorts of ways.

5. Enable Slide to Shutdown

This trick is complicated and probably not worth the effort for what you get out of it, but you can use it to slide your computer to the off position. Right-click on the desktop and select New > Shortcut. In the ensuing pop-up window, paste the following line of code:

%windir%\System32\SlideToShutDown.exe

This creates a clickable icon on your desktop, which you can rename. Right-click the file and enter Properties to add a shortcut key or double-click the file to run the program. This prompts a pull-down shade to appear, which you can drag with the mouse down to the bottom of the screen. Keep in mind; this is shutdown, not sleep.

6. Drag to Pin Windows

This feature was available as far back as Windows 7 but has some extras in Windows 10. Grab any window and drag it to the side, where it will "fit" to half the screen. You also have the option of dragging the window to any corner to have the window take over a quarter of the screen instead of half.

If you're using multiple screens, drag to a border corner and wait for a prompt signal to let you know if the window will open in that corner. You can prompt similar behavior by using the Windows key plus any of the directional arrow buttons.

7. Quickly Jump Between Virtual Desktops

Do you like to multitask on your PC? In Windows 10, Microsoft finally provided out-of-the-box access to virtual desktops. So now you can *really* multitask.

To try it out, click on Task View (the icon next to the search box). This will separate all your open windows and apps into icons. You can then drag any of them over to where it says "New desktop," which creates a new virtual desktop. This would allow you to, say, separate your work apps, personal apps, and social media into different desktops.

Once you click out of Task View, you can toggle between virtual desktops by pressing the Windows key + Ctrl + right/left arrows. To remove the virtual desktops, just go back into task view and delete the individual virtual desktops—this will not close out the apps contained within that desktop, but rather just send them to the next lower desktop.

While you're here, you should notice that Windows saves a timeline on all your app activity on this page. You can save up to 30 days of activity when signed in with a Microsoft Account. Click on an activity and open it back like just like the day you were using it.

8. Customize the Command Prompt

This feature will probably only be useful to a narrow niche of users, but if you like to dig your virtual fingers into the innards of Windows via the Command Prompt, Windows 10 provides a few customization options.

To access the Command Prompt interface in Windows 10, click on the Windows menu and type "Command Prompt" to bring up quick access to the desktop app. Click the icon to open the Command Prompt, then right-click at the top of the window and choose Properties.

This pop-up window allows you to personalize the experience by changing the font, layout, colors, and more of the Command Prompt. You can also turn the window transparent by opening the Colors tab and moving the Opacity slider. This feature lets you code away in the Command Prompt while simultaneously observing the desktop.

9. Nearby Sharing

In an open document or photo, you can share the file directly with nearby devices the same way Apple's Airdrop works. Click the Share icon atop your doc or photo toolbar to open the panel, and then click Turn on nearby Sharing to see which nearby recipients are in range.

Control this feature by going into Settings > System > Shared Experiences to turn Nearby Sharing on and off. You can also set it to share with anyone or only your devices for easy file transfer.

10. Dark Mode and Light Mode

Windows 10 gives you a significant amount of control over color themes. Open Settings > Personalization > Colors and you can set the operating system to either dark mode or light mode. These themes change the color of the Start menu, taskbar, action center, File Explorer, settings menus, and any other programs that are compliant with these palette changes.

There is also a custom option that will let you set one theme for Windows menus and another for apps. Want a little more color? There are swatches of color themes available to choose from that can help your menus and taskbars really pop.

Keyboard Tricks to Save Time Every Day

Using a computer mouse is all well and good but wouldn't you like to save yourself some time? Check out these twelve keyboard tricks to help you work more efficiently.

1. Ctrl + C & Ctrl + V

If you don't know how to copy and paste using your keyboard, it's likely that you're not going to know any of the other keyboard shortcuts featured on this list either. At least you'll find it useful!

2. Windows + L

Do you trust your co-workers not to access your computer and send an embarrassing email while you're away from your desk? If not, hit the Windows key + L key and you'll be able to lock your computer, preventing unauthorized access!

3. Windows + Left & Windows + Right

Want to look at two windows simultaneously? Windows 7 makes it easy. All you need to do is hit the Windows + Right Arrow key to snap a window to the right and Windows + Left Arrow key to snap a window to the left. Windows + Up Arrow will maximize a window and Windows + Down Arrow will minimize it. This shortcut also works in the desktop view of Windows 8.

4. Ctrl + Shift + V

Want to copy some text into your email or online document without the funky font and garish colors? Simply hit CTRL + Shift + V on your keyboard and the text will be pasted as plain text. This works in both Chrome and Firefox.

5. Ctrl + D

Like what you are reading? Hit CTRL + D to save this page to your bookmarks! It works on all other web pages too, providing you're using Chrome, Firefox or Internet Explorer as your browser.

6. Alt+ Left & Alt + Right

How much time do you spend clicking the back and forward button on your web browser every day? Wouldn't it be great if there were a more convenient way to do it? Oh wait there is! Simply hit ALT + Left Arrow or ALT & Right Arrow on your keyboard to flick back and forth between web pages.

7. Ctrl + S

When you're working on a document in Word or any other program, it's recommended that you keep pressing CTRL + S every few minutes. This will save any changes you've made to your documents, so if your computer crashes, you won't lose your work.

8. Shift + Space

If you're editing a spreadsheet in Microsoft Excel and Google Docs and want to highlight an entire row without using your mouse, simply hit Shift + Space. Highlight other rows by hitting CTRL + Shift + Up Arrow or CTRL + Shift + Down Arrow.

9. Ctrl + Z

Deleted something in your document by accident? Simply press CTRL + Z and it will undo the last change you made. Press it multiple times to undo multiple changes. If you want to redo the undo then press CTRL + Y. Are you still with us?

10. Ctrl + F

This is one of our favorite keyboard shortcuts as it saves so much time! If you're looking for a particular word in a document, email or on a web page, simply press CTRL + F and it will open Find in any program. Type in your word and it will highlight it throughout the text, making it easier for you to spot.

11. Ctrl + Home & Ctrl + End

CTRL + Home will move your cursor to the beginning of your document or press CTRL + End to move your cursor to the end of the document. This should also work for web pages too.

12. Ctrl + P

Want to print this document so you can stick it by your computer and learn your shortcuts? Simply hit CTRL + P to bring up your printing options!

How to Make Symbols with Keyboard

***These shortcuts are mainly works in text editor.**

Alt+0153	™-Trademark symbol
Alt+0161	¡ upside down exclamation sign
Alt+0169	©Copyright
Alt+0174	® Registered trademark symbol
Alt+0176	° Degree symbol
Alt+0177	±plus-minus sign
Alt+0178	2 as symbols or text
Alt+0179	3 as symbols or text
Alt+0180	´ as symbol
Alt+0181	µ sign
Alt+0182/20	¶-paragraph mark
Alt+0183	·(.dot) symbol
Alt+0184	¸(comma) symbol

Alt+0190	¾ fraction, three-fourth
Alt+0191	¿ upside down question mark
Alt+1	☺ smiley face
Alt+2	☻ black smiley face
Alt+3	♥ heart shape smiley
Alt+4	♦ Diamond symbol
Alt+5	♣ Club suit Emoji
Alt+6	♠Spade Emoji
Alt+7	• filled circle bullet
Alt+8	◘ Reverse bullet
Alt+9	○ unfilled circle bullet
Alt+10	◙ Reverse bullet
Alt+11	♂ Male sign
Alt+12	♀ Female sign
Alt+13	♪ Eighth note music symbol
Alt+14	♫ Beamed pair music symbol
Alt+15	☼ Sun symbol
Alt+16	► Triangle bullet
Alt+17	◄ Reverse side triangle bullet

Alt+18	↕ Up-Down double headed arrow
Alt+19	‼ Double exclamatory sign
Alt+8721	∑ Auto sum sign
Alt+ 251	√ square root sign
Alt+ 236	∞ Infinity sign
Alt+24	↑ up arrow
Alt+25	↓ down arrow
Alt+26	→ right direction arrow
Alt+27	← left direction arrow
Alt+29	↔ left-right double headed arrow
Alt+30	▲Triangle bullet
Alt+31	▼ downward direction triangle
Alt+33	! Exclamatory sign
Alt+34	"double inverted comma
Alt+35	# Hash tag
Alt+36	$ Dollar
Alt+37	% percentage sign
Alt+38	& (and/ampersand)
Alt+39	'single inverted comma

Alt+40	(Left parenthesis sign
Alt+41	) Right parenthesis sign
Alt+42	* (Star) sign
Alt+43	+ (Plus) sign
Alt+44	, (Comma) sign
Alt+45	- (Minus) sign
Alt+46	. (.dot) sign
Alt+47	/ (slash) forward

Facts about Passwords That Will Make You THINK

Fact 1: Passwords are easily hacked because most humans follow similar patterns

At the beginning of the Web and when passwords were first used, the most popular password was '12345'. Today, it may be longer, but is hardly safer - '123456'. Additionally, research has found that women are famous for using personal names in their passwords, and men opt for their hobbies.

Fact 2: 59% of people use the same password everywhere

91% of people know that password recycling poses huge security risks, yet 59% continue to use the same password everywhere. Therefore, if a hacker was to crack one password, they would be able to gain access to all other accounts!

Businesses should ensure they pay close attention to employee password hygiene. Studies have shown that there can often be a lap over with the passwords created for personal and work accounts; 62% of people use the same password for work and personal accounts.

Password generators are great if you struggle to come up with multiple, strong passwords. They are tools that will automatically generate a password using parameters such as mixed-case letters, symbols, numbers, length and strength.

Fact 3: 7 in 10 people no longer trust passwords to protect their online accounts

Passwords are required for nearly everything we do online. So, if people no longer trust them, what is the answer?

Multi-factor authentication (MFA) or two-factor authentication (2FA) is authentication methods that verify a user's identity by requiring multiple credentials. These include something you *know*, something you *have* and something you *are*. Something you know could be a password, something you have is a possession such as a generated code on your phone, and something you are could be facial recognition, a fingerprint or an eye scan.

As traditional usernames and passwords can be stolen, they have quickly become a target of hackers. This explains the lack of trust in them for many. MFA or 2FA are effective ways to provide enhanced security for all online accounts.

Fact 4: 86% of people who use 2FA feel their accounts are more secure

Ever since the start of the digital revolution, passwords have been the mainstream form of authentication. Unfortunately, as passwords and encryption methods have become more complex, so have the skills of hackers.

2FA is an essential element of cyber security that all businesses should implement as it adds that extra layer needed to immediately neutralize the risks associated with compromised passwords. Implementing it can be done with relatively little pain for users, and usually, with little or no expense to your organization.

At entrust IT we understand the importance of good cyber security practices, which is why with our Hosted Desktop, Hosted Application and Office 365 products, we encourage the use of 2FA. This is especially true for more demanding environments such as legal, financial services and local government where it is strictly enforced.

Fact 5: 90% of passwords can be cracked in less than six hours

Think you have a strong password? Think again...

Hackers are continuing to become more sophisticated and have a variety of ways in which they can crack your passwords to gain access to your online accounts. One way to help keep secure is to understand the methods they use, here are four:

1. **Dictionary attack** - A dictionary attack is a method that systematically enters words that can be found in a dictionary. Hence, the name. The only reasons this kind of attack works is because users are remaining to rely on easy-to-guess words for their passwords.
2. **Brute-Force attack** - A brute-force attack is when hackers have software that tries to guess every possible combination until it hits yours. They often begin with the most commonly used passwords first and then move onto more complicated phrases.
3. **Credential stuffing** - Credential stuffing proves the dangers of re-using usernames and passwords for numerous accounts. It works where credentials obtained from a data breach on one platform are used to attempt log INS on other platforms.
4. **Social engineering**- Phishing has remained one of the top social engineering methods used by hackers to crack passwords. They do this by appearing as a trusted source and concoct a scenario for handing over login credentials or other sensitive personal data.

Fact 6: 18% of employees share their passwords with others

Password sharing is a common mistake of many and can seriously compromise an organization's cyber security. But why do employees do this? Research has shown 42% of workers say they do it to more easily collaborate with team members, as well as 38% saying they share passwords because it is company policy.

If a hacker gains entry to your system, shared passwords will make it much easier for them to access other parts of the network. Additionally, how do you establish exactly who is doing what? By taking the time to put an updated password policy in place, you can minimize the risk of both internal and external threats related to password sharing.

Typing Skills

Here are some tips which always help you to keep your proper typing posture, way to typing right in manner.

Proper Typing Posture

- Keeping your feet flat on the ground and your neck and back straight.
- Adjusting your elbows to an angle between 90 and 110 degrees.
- Keeping your wrists in a neutral stance.
- Moving your monitor so that the top of your screen is at eye-level.
- Adding appropriate ergonomic **typing** accessories.

Typing speed

- Do not rush when you just started learning. Speed up only when your fingers hit the right keys out of habit.
- Take your time when **typing** to avoid mistakes. The speed will pick up as you progress.
- Always scan the text a word or two in advance.

Keep these important tips in mind while typing.

1. Improve Your Workspace.
2. Fix Your Posture.
3. Hold Your Posture.
4. Familiarize Yourself with the Keyboard.
5. Close Your Eyes and Say the Keys out Loud as You Press Them.
6. Start Slowly With Touch-Typing.
7. Don't Look at Your Hands.
8. Practice, Practice, Practice..................

Beginner exercises to improve typing skills

Once you're sitting in the proper position you can use the exercises below. Follow these directions:

- Type the words in the exercises below with a single space between them and after punctuation.
- Type the passage as you see it with the line breaks.
- Go slow and be deliberate.
- See how many words you can type in a row without looking at your hands.
- See how many words you can type without making any mistakes.
- Note which fingers, letter keys, or words give you any trouble. Pay attention to your form so you don't get fatigued or hurt yourself, and take breaks where you stand and stretch.

Video Conferencing App's

- **Zoom:-** What has been the go-to video conferencing app of 2020, Zoom shot to fame ever since the lockdown kicked in. The app is available for download for free on Android, iOS and Microsoft but you can always upgrade your app to avail the premium services that Zoom offers.

The app has been able to garner more than 300 million users over the course of the past couple of months and is one of the most-used apps today.

With the free version of the app, users get 40-minutes of uninterrupted video calling with an option to record the call.

The paid version of the app offers features likes a maximum of 100 participants, unlimited group meetings, social media streaming and 1GB cloud recording (per account).

What makes Zoom an attractive proposition is the interface is relatively simple to use and it also offers different backgrounds and audio enhancement features.

- **Microsoft Teams:** Microsoft Teams is the first name that comes to your mind when one is talking about a video conferencing solution for an organization. Since the platform is closely integrated with the rest of Microsoft's suite of apps users do not have any trouble sharing work on the platform during meetings.

In terms of subscription plans, Microsoft offers its Microsoft 365 Business Basic subscription for Rs 125 per month where you also get Teams loaded with other Microsoft apps. The Business Standard plan costs Rs 660 while the complete Office 365 E3 plan costs Rs 1,320 per month.

With Teams, you get the ability to host up to 250 members with screen sharing and call recording bundled. You can also connect it the calendar which can help you schedule meetings in advance. In comparison to its competitors, it also offers an extensive array of exclusive features like Calls and Activity among others.

- **Skype.** ... Before Zoom, Teams and other video apps, people usually turned to Skype as the default video conferencing app. What was a surprise for many is that Skype did not advertise itself more aggressively during the lockdown which is where it lost out on a user base that found Zoom and Teams as a better option.

Initially, the app was created to make voice calls over the internet but later introduced a video conference feature. It is compatible with Windows, Mac, Android and iOS operating systems. The app can support up to 50 participants in a single call and also offers other features like screen sharing and also sharing documents.

It also integrates other Microsoft apps like Word, OneDrive and Outlook as well.

The paid version of the Microsoft-owned platform is available for a starting price of Rs 147 per month.

- **Google Meet:** Google Meet earlier known as Google Hangouts Meets is Google's challenger in the video conferencing space. The app is available in Android and iOS devices and can also be accessed on the Chrome browser.

Like its competitors, it also presents its participants in a tiled format with up to 16 participants in a single window. It supports 2-step verification for security and also offers screen sharing and real-time captions to its users.

Since Meets is an integral part of the Google Ecosystem, the search giant has been pushing its video conferencing platform a lot. The best part about Meet is that it offers a minimum 60-minute limit for free and can accommodate up to 100 participants at a time.

For users who want to avail the premium features of the app like live streaming for more than 100,000 views, inviting 250 participants and option to record and save on Google, they have to pay a minimum of Rs 125 per month.

- **Cisco WebEx**: Not a very commonly known video conferencing platform but not one you'd want to count out of the equation. The Cisco WebEx video conferencing app provides a comprehensive solution in terms of video and chat as well.

It comes with HD video conferencing features bundled with compatibility across various platforms like Windows, iOS and even Android. It also allows users to share documents across the platform which is a key feature when it comes to organizational needs.

You can use the service for free for a period of 30 days with limitations in terms of video conferencing duration among others. The premium subscription of the app starts at Rs 945 per month in which you can accommodate up to 50 participants, get 5GB of cloud storage and also get 24×7 customer support service.

Full Forms

COMPUTER	→	Common Operating Machine Purposely Used for Technical and Educational Research.
AI	→	Artificial Intelligence.
USB	→	Universal Serial Bus.
KEYBOARD	→	Keys Electronic Yet Board Operating A to z Response Directly.
MOUSE	→	Manually Operated User Selection Equipment.
WIFI	→	Wireless Fidelity.
INTEL	→	Integrated Electronics.
DELL	→	Digital Electronic Link Library.
HP	→	Hewlett Packard.
ACER	→	Australian Council for Educational Research.
BIOS	→	Basic Input/output System.
LCD	→	Liquid Crystal Display.

LED	→	Light Emitting Diode.
TFT	→	Thin Film Transistor (Type of LCD flat panel display).
C Language	→	Compiler.
C++ Language	→	Advanced Complier.
PDF	→	Portable Document Format.
JPEG	→	Joint Photographic Experts Group (image file extension).
PNG	→	Portable Network Graphics (image file extension).
GIF	→	Graphics Interchange Format (image file extension).
TIFF	→	Tagged Image File (image file extension).
PSD	→	Photoshop Document.
EPS	→	Encapsulated Postscript.
WebM	→	WEB Media (video file extension).
MKV	→	Matroska Multimedia Container (video file extension).

FLV	→	Flash Video (video file extension).
Nvidia	→	(Graphics card maker).
VGA CABLE/PORT	→	Video Graphics Array.
HDMI CABLE/PORT	→	High Definition Multimedia Interface.
MONITOR	→	Mass on Newton Is Train on Rat.
AIRTEL	→	Affectionate Interested Respectful Tolerant Energetic and Loving.
JIO	→	Joint Implementation Opportunity.
LG	→	Lucky Gold star.
INTEX	→	Imaginative Nurturing Trusting Excitable Xenial Advertisement.
BROTHER (BRAND)	→	Brilliant Role model Orthodox Teacher Horrible Enchanting Roommate.
INFOSYS	→	Information Systems.
HCL COMPUTERS	→	Hindustan Computers Limited.
TCS (COMPANY)	→	Tata consultancy Services.

WIPRO(COMPANY)	→	Western India Products.
NEWSPAPER	→	North East West South Past And Present Events/Everyday Report.
BOOK	→	Big Ocean of Knowledge.
MOBILE	→	Modified Operation Byte Integration Limited Energy.
GOOGLE	→	Global Organization of Oriented Group Language of Earth.
YAHOO	→	Yet Another Hierarchically Organized Oracle.
FTP	→	File transfer protocol.
HTTP	→	Hypertext Transfer Protocol.
VIVO	→	Video In Video Out.
OPPO	→	Obedient Punctual Perceptive Outstanding.
TEACHER	→	Talented, Elegant, Awesome, Charming, Helpful, Efficient, Receptive.
ASCII	→	American Standard Code

		for Information Interchange.
DFD	→	Data Flow Diagram
NOS	→	Network Operating System.
DLL	→	Dynamic Link Library
MSI	→	Microsoft Installer
TPS	→	Transaction Per Second
GP	→	Graphics Port
CGI	→	Common Gateway Interface.
MMU	→	Memory Management Unit.
BCR	→	Bar Code Reader
YB	→	Yotta Byte
EB	→	Exa Byte
ZB	→	Zetta Byte
Mac	→	Macintosh
DMI	→	Desktop Management Interface.
NTP	→	Network Time Protocol
VT	→	Video Terminal
D2D	→	Device to Device

FDC	→	Floppy Disk Controller
NFS	→	Network File System
SFC	→	System File Checker
AMD	→	Advanced Micro Devices
BAL	→	Basic Assembly Language.
CDN	→	Content Delivery Network.
RJ45	→	Registered Jack 45
OOPS	→	Object Oriented Programming System.
SRAM	→	Static Random-Access Memory.
VRAM	→	Video Random Access Memory.
BPS	→	Bits Per Second.
LPX	→	Low Profile Extension.
MUI	→	Multilingual User Interface.
MFD	→	Multi Function Device
MBR	→	Master Boot Record
BCD	→	Binary Coded Decimal
CAM	→	Computer Aided Manufacturing.

DFS	→	Distributed File System.
WMA	→	Windows Media Audio
WIFI	→	Wireless Fidelity
VFS	→	Virtual File System
SCSI	→	Small Computer System Interface.
LTE	→	Long Term Evolution.
MPEG	→	Moving Picture Experts Graphics.

The author found from his 10 years of experience that students have a great desire to acquire knowledge of Computers/IT. As all know that computer has become very important in today's technological age. Computer is one of the household gadgets in millions of houses around the globe. The impact of computers can be felt almost in all fields of human activities. This book is framed for those who wish to spread their wings in Computer/IT. This book will be helpful for those also who are pursuing academic courses on the subject. Even though they are reading academic books for this, they still need to learn the techniques of it briskly, the author's only dream is for every learner should adopt the latest technologies and would be an engineer in his own right. So this effort has been made by the author to make them learning these techniques/skills.

www.ingramcontent.com/pod-product-compliance
Ingram Content Group UK Ltd.
Pitfield, Milton Keynes, MK11 3LW, UK
UKHW021935190726
13853UKWH00004B/1457

9 789354 728570